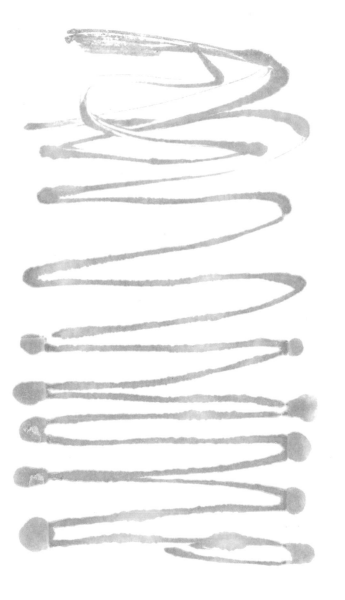

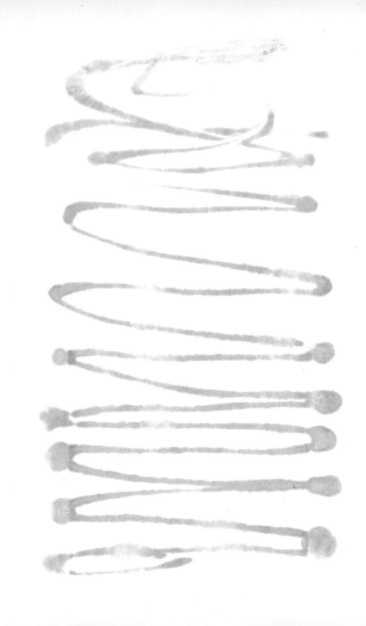

CRAFT AND ART

ART FOUNDRY

CRAFT AND ART

Christian Hauser

ART FOUNDRY

VAN NOSTRAND REINHOLD COMPANY
New York · Cincinnati · Toronto · London · Melbourne

Library of Congress Catalog Card No.: 73-8467
ISBN 0-442-29990-7

Printed in Switzerland.

Published in 1974 by Van Nostrand Reinhold Company Inc., 450 West 33rd Street, New York N.Y. 10001 and Van Nostrand Reinhold Company Ltd., 25–28 Buckingham Gate, London SW1E 6LQ.

Van Nostrand Reinhold Company Regional Offices:
New York, Cincinnati, Chicago, Millbrae, Dallas.
Van Nostrand Reinhold Company International Offices:
London, Toronto, Melbourne.

CONTENTS

Metal casting in an alchemist's laboratory. 1535.

1 INTRODUCTION

This third volume in the series "Craft and Art" will enable the reader to discover one of the oldest crafts in the world: metal casting, and more particularly the casting of art objects. Men have been working metal for utilitarian or ornamental purposes ever since the Bronze Age; from those remote times onwards metal-working techniques, specially casting, have never ceased to develop.

The principle of casting probably stems originally from sculpted metal images, usually masks, obtained by applying thin sheets of metal to a carved wooden form covered with a layer of mineral tar. When a perfect replica of the form had been worked out the metal was chiselled, polished and burnished. Certain Sumerian and Egyptian works, of which the magnificent gold mask of the pharaoh Tutankhamen is an outstanding example, were fashioned in this way.

This technique was painfully slow and delicate; not infrequently the form had to be made two or three times over. The discovery of casting greatly simplified the work, as it called for one mould only and resulted in a much thicker metal object

which was easier to retouch. The conversion to casting, however, was a gradual process. Very likely the first small cast objects were used to build up images produced by the older technique. The precise date at which casting was introduced is still unclear, but the oldest known bronze cast is Indian and was made about five thousand years before our era.

Since then generations of founders have succeeded one another without in fact introducing any major changes to the art of casting in general, or to the lost wax process in particular. Lost wax casting is basically very simple: the object to be reproduced in metal is either modelled in *wax,* if small and solid, or in the case of a large hollow work, fashioned in wax around a *core* of refractory material. This wax model is itself surrounded by a *mould* of refractory material, called the *investment* or *ludo mould,* which must be plastic enough to take a perfect imprint of the wax. The

Henri Presset, La Femme d'Arve, *1967. Bronze.*

Small winged lion, *Iron Age. Urartu bronze.*

8

9

whole is then heated to a very high temperature until the wax disappears completely forming a cavity into which the molten metal is poured replacing the lost wax.

A description of the practical applications of this and one or two other processes used for casting takes up the major part of this volume. May it tempt you to spend a day in a foundry, enthralled by the weird spectacle of the casting operations.

Pair of rein-guides, *Iron Age. Bronze from northern Iran.*

2 MASTERS OF FIRE

Before visiting the foundry itself however we must study briefly the production of the basic material – metal – which played a big role, sometimes the leading role, in the development of ancient civilizations.

Founders, like potters before them, are "masters of fire". Fire can change matter from one state to another. This fact seems elementary and obvious today, but for centuries, because of it, the "metal man" was elevated to the rank of demigod or held in awe as a being endowed with

Stag, *Iron Age. Amlash bronze.*

divine or magic powers. Only such powers could explain the ability to modify a part of the "world" by fire, not only by accelerating a natural process of transformation but also and chiefly by creating something new out of what could be found in nature.

This transmutation was inevitably accompanied by a complicated ritual as is witnessed by this Assyrian manuscript:

"When you set out the plan of an ore furnace, you will choose a favourable day in a favourable month, and thereupon you will set out the plan of the ore furnace. While they build the furnace you will watch (them) and labour yourself (in the furnace house): you will place embryos (Ku-bu) in the furnace chapel, no stranger may enter, nor impure person walk before them: you will offer up to them due libations; the day that you place ore in the furnace, you will perform a sacrifice before the embryo; you will place there a censer with incense of pine, you will pour beer before them.

"You will light a fire under the furnace and place the ore in the furnace. The men you have chosen to attend to the furnace

The moulds are piled up. Work starts on building the furnace walls.

must purify themselves, upon which you will establish them in care of the furnace. The wood you will burn under the furnace will be styrax, large thick logs, stripped of bark, which have not been exposed in faggots but conserved under envelopes of skin cut in the month of Ab. This wood will be placed under your furnace."

This translation of the word "Ku-bu" is disputed, but all versions imply some sort of embryological significance showing that a founder's work was considered as an acceleration of the creation of metals which were supposed to "grow" inside the earth (in its womb) and there become transformed. This last idea is still current in certain parts of South-East Asia, where it is believed that a bronze object left sufficiently long under ground will change into gold. Until recently the notion of the sacredness of the act of casting was also evident in some regions of black Africa, where founders set up camp in the immediate vicinity of the mines and lived in complete continence during the whole smelting period – sometimes more than six months – allowing no woman near the furnaces.

All these taboos and ritual acts which accompanied the transformation of ore into metal are evidence of the fear and respect with which ancient craftsmen accomplished what they felt was the Earth-Mother's work when they accelerated and perfected the "growth" of an ore by transplanting it in a sort of "artificial womb", the furnace.

As was pointed out above, the "master of fire" occupied a special niche in society, not unlike that of a witch doctor – or, although much more rarely, of a pariah. Few and far between during the Bronze Age, founders grew in numbers and influence with the advent of the Iron Age, when, having become iron-smiths, they were able to fashion everyday objects in iron the use of which became widespread. Bronze and copper, however, remained rarer material and were used primarily for ornamental purposes.

Hoggenberg, costume and instruments of ore miners in the Harz mines, Germany, early 17th century. Hand-coloured engraving.

3 THE METAL

Since antiquity the alloy bronze has been the metal most often used for casting statues. Silver and gold have been used much less frequently, and, like lead, are mainly reserved for small scale work. Nowadays some artists try out more modern metals such as steel and aluminium.

Bronze is composed principally of copper and tin, the proportions being roughly 90% copper to 10% tin.

Copper on its own is very soft and does not pour easily when molten. So the tin is added to provide both strength and fluidity. Immediately after casting this alloy bears no resemblance to "bronze" but looks more like reddish brass. The action of chemical agents, some of which are in suspension in the atmosphere, can give to a bronze surface a variety of hues ranging from brown to blue through a whole scale of greens.

Almost from the beginning other metals, usually zinc and lead, have been alloyed with copper either in place of tin or in addition to it. The use of these metals can be explained by both the difficulty of obtaining tin and their chemical properties. The tin used by the Greeks had to cross the whole continent from the British Isles where it was mined. Lead was introduced very early on, mostly in coins, whereas zinc only appears during the Roman period; it then sometimes replaced tin completely, particularly in bronze plate. From then on these metals were retained because of the ways in which they improved the alloy. Zinc, for instance, reduces the retention of gases during casting, which is a good thing as these gases weaken the metal and give it an unattractive porous surface. As to lead, it lowers the melting point, makes the alloy more fluid and facilitates retouching work once the metal is cool.

The proportions in which these metals are found vary considerably depending on the alloy and its use. A certain number of known alloys can be seen in the table on page 20.

Greek mirror bronze needed to be very light-coloured, but it was not necessary for it to be as tough as bronze used for statues, so the founders added lead to the alloy.

Giorgo Vasari, 16th century painter, architect and art-historian, the author of *Lives of the best painters, sculptors and architects,* provides us with the first formula for Italian Renaissance bronze, while the second is from the pen of Pomponius Gauricus, whose *De sculptura* was published in 1504.

As for the French formula it is the one and only one used by two famous brothers, the founders Jean-Jacques and Jean Balthasar Keller, natives of Zurich who lived in Paris and were responsible for most of the statues in the Versailles gardens, cast during the second half of the 17th century.

The various differences between the alloys in the following table affect the weight of the statues and the colour of the metal. Italian works are lighter and much redder than French statues of the same period.

Metallurgy, copper works.

The last three columns of the table show that even today there is no absolutely perfect composition for the alloy. Every founder uses his own formula which is often a jealously guarded secret. These alloys may in addition contain other components such as phosphorous and aluminium.

Only bell-founders are restricted to casting with alloys of a constant composition which varies little from the formula 78% copper to 22% tin.

The melting point for these various mixtures is reached at around 1010° C and varies according to the amount of white metal present in the alloy.

The mechanical properties of the alloy depend mainly on its cooling down. The crystaline structure is at its best and the metal strongest when the freezing point is reached rapidly. Also, an excessive temperature increases the production of gases, and the slower the metal cools the more its surface is attacked by gas. To combat

The founder introduces lumps of bronze into the furnace.

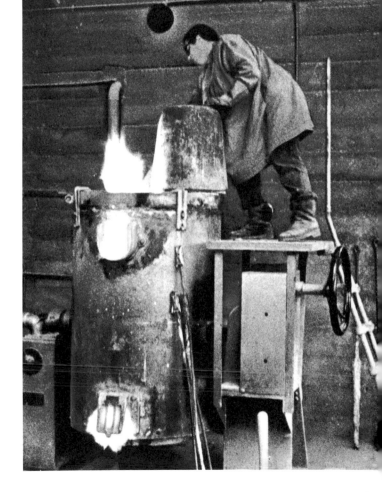

	Greek Bronze Mirror	Greek Bronze Statues	Chinese or Japanese Bronze	Italian Bronze Renaissance I	Italian Bronze Renaissance II	French Bronze XVII	Modern Bronze I	Modern Bronze II	Modern Bronze III
Copper	62 to 71%	88%	80%	75%	86%	90%	90%	85%	85%
Tin	23 to 32%	6 to 9%	4%	25%	12%	2%	6.5 to 7%	5%	15%
Lead	5 to 7%		10%		2%	1%	3 to 3.5%	5%	
Zinc			2%			7%		5%	
Other metals		traces	4% iron						

these gases the founder adds to the mixture oxygen-releasing chemical compounds which counteract hydrogen and carbon monoxide. This, however, gives rise to another danger as excess oxygen can cause the oxidization of some parts of the alloy, so phosphor-copper is added: the phosphorous contained in the copper combines with the oxygen to form a precipitate which is removed before pouring.

Metallurgy, Chinese plate end of 18th century. Drawing and water colour.

4 THE ARTIST AND HIS FOUNDER

It is rare, even exceptional, for an artist to be able to cast his own work. He must, then, leave this job to his founder, although he will collaborate with him on a large number of operations. This has long been the custom; Ghiberti who had a foundry of his own, or Cellini, who could cast his own work because of his superior technical knowledge, were considered exceptional even in their time.

While the artist's presence and his collaboration are usually required at the beginning, and sometimes at the final stages of casting, the professional craftsman has the sole responsibility of the central part of the operation.

A sculptor wishing to prepare a work for casting has the choice of several methods which vary not only in relation to the type of casting, lost wax or sand, but according to his intentions.

If he chooses the lost wax process he has two possibilities: he can either make a unique model which is destroyed during casting or use the hollow cast method in

Alberto Giacometti in his studio working on a plaster.

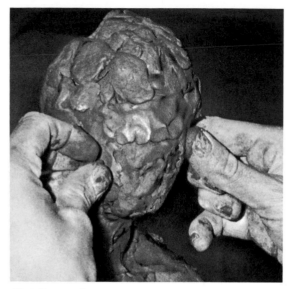

Modelling in clay.

which the master cast is preserved and can be used again.

The main particularity of the first method is that the original model, in other words the positive, is itself melted, so the wax is not poured into a previously prepared mould. The model can only be used once, it is destroyed during casting. There are two ways of setting about this method,

the sculptor can make the model in either wax or clay. For small objects and even certain masks – they must be very thin – he can work directly in wax which will be both the core of the work and its surface. The result will be in solid bronze; this creates certain problems of weight and cooling which are minimized if the work is small or very thin.

It is much more common for an artist to model in clay on a rigid armature the form that will become the core of his sculpture once dried and baked. A layer of wax, which will later be replaced by bronze, is applied to the perfect core modelled previously. The artist can remodel the layer of wax provided he makes sure its thickness remains constant wherever possible. Obviously the clay original cannot be used again; it is usually removed to lighten the bronze. Classical Greek statues were generally made this way.

We must thank Japanese, and later, Renaissance founders for perfecting the second process which makes it possible to preserve the master cast. The latter can be made in almost any material: clay, stone, plaster, bronze, wood etc. ... The sculptor is thus

able to have bronzes made from work
carried out in the material of his choice
which need not be destroyed. He must
then take a negative impression of his
work, a mould, which will be lined with
wax and given a core which corresponds
roughly to the form of the master cast.

More often than not contemporary
artists choose this process; not only does
it prevent damage to the master cast but
it dispenses with the relatively complicated
task of modelling wax and makes it possible to cast several examples of the same
sculpture.

When a sculptor chooses the sand process for casting his work, he must provide
a definitive model from which an impression is taken in sand using a two-piece
mould. Overcuts are treated separately.

The sculptor only becomes master of
his work again when the mould is opened
after casting, he can then leave everything
to the founder including the final polishing
or, on the contrary, take over the work
himself.

*First negative plaster mould taken from the original
model.*

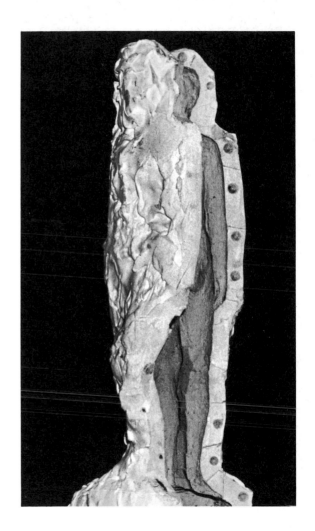

Médardo Rosso: wax.

26

5 THE LOST WAX PROCESS

The master cast

An artist who wishes to have a bronze made by hollow casting hands the original over to his founder; as we have seen it can be made of almost any material. The founder prepares the wax which will later be replaced by bronze and pours it into a negative mould which can be made of various materials. If the master cast is in clay, plaster provides the best mould; however, flexible materials are often used; they make it unnecessary to divide the mould into more than two pieces. Gelatine, polyvinyl chloride (P.V.C.) and other plastic materials containing latex produce the best results, and for really perfect work it is wise to choose *gelatine* although this restricts the number of waxes that can be cast. P.V.C. is extremely durable and elastic but expensive and fairly unpleasant to handle. Latex is also long lasting but polymerizes quickly on contact with air.

The technique for preparing a mould is the same for all these materials; first the surface of the master cast is sealed for protection, then a platform of clay is built round it level with the proposed seam. The half of the master cast above the plat-

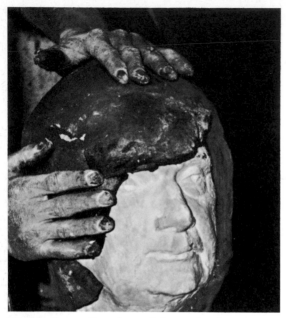

The founder applies a layer of clay to the plaster master cast.

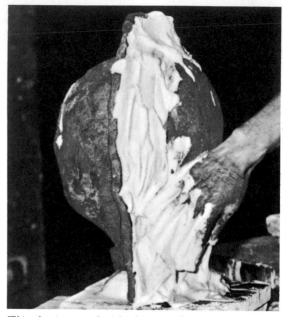

This clay is covered with plaster to form the first half of the outside shell.

form is covered with a layer of clay topped by a shell of plaster. When the plaster is dry the whole thing is turned over, the clay platform removed, and the second half of the plaster shell built up as before over a layer of clay, meeting the greased edges of the first half at the seam. The clay

lining is now removed from one side of the master cast leaving an empty space between the cast and the plaster shell into which is poured gelatine heated in a double boiler. The shell is opened again, the clay removed from the second side of the master cast and the set gelatine and plaster

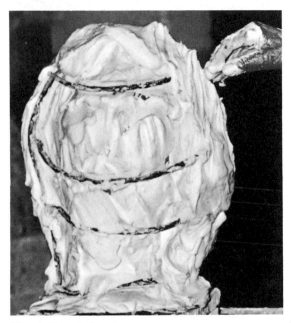

The shell is reinforced.

When the plaster shell has been emptied of clay and carefully cleaned, the master cast is put back in place before the gelatine is poured.

surfaces well greased before the two halves of the shell are once more joined and the second half of the gelatine mould is poured. The shell is then opened and the master cast is extricated with care from the gela-

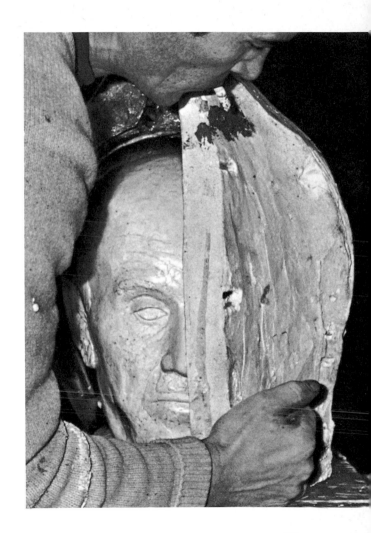

tine mould. Wax can now be poured into the space left by the master cast.

Plaster and clay can also be used as moulds for wax if the necessary precautions are taken to facilitate the removal of the wax positive.

The wax and its preparation
In foundry vocabulary the term "wax" applies to both the material and the positive cast made with it. The wax used for casting must be strong enough not to become deformed by frequent handling but not so hard that it is difficult to retouch. Beeswax has been preferred for a very long time, but today its price and scarcity favour synthetic materials with which it is sometimes mixed.

To bring wax to melting point it is best to use a double boiler which heats the whole mass regularly without the danger of an explosion. It is useful to prepare two containers for the melted wax, one for brush work, the other for pouring: these

The plaster shell is reshut and the founder fills it with gelatine.

Half of an open mould : the solidified gelatine mould can be seen between the outside plaster shell and the master cast.

two waxes are often different in quality. In fact the operation has to be carried out in two stages: first a thin film of hot wax is painted onto the inside surface of both halves of the mould, these are then assembled and the necessary quantity of wax is poured in, all at once or in several goes, until the desired thickness has been obtained.

Before brushing on the first coat of wax it is necessary to place small ventilation tubes in position to prevent the formation of air traps which could cause parts of the wax positive to collapse: for the wax, being hot and soft still, could then easily become unstuck under its own weight.

Two or three coatings of very hot wax must be applied by brush with great care so that they adhere perfectly to the mould with no cracks or air bubbles.

The wax which is then poured into the firmly re-assembled mould should be at a temperature near its setting point in order not to damage the first coating and to freeze quickly. The mass of hot wax is poured into the mould, left a few seconds, and then poured out again, all except a film about 1/8 in. (3 mm.) thick which is deposited on the inside walls of the mould. This operation is repeated until the desired thickness has been obtained all over. On a mould that is still warm little wax is deposited. The thickness of the layer can be doubled by waiting for the mould to cool. Depending on the quality of the wax this operation is done in one or in several goes.

If the mould is entirely made of plaster and relatively small it can be immersed completely in very hot water, then removed, filled with hot wax and re-immersed in cold water up to its opening, filled with liquid wax, which must stay clear. After a few seconds the surplus wax can be poured out. For large scale work the founder applies the wax by hand using a mixture which softens at skin temperature. Thin strips of wax are applied to a coating previously painted onto the mould with a brush.

The wax is now ready, but before *runners* and *risers* can be put in place and the final

Joannis Avramidis:
1) Model for a column *A V, 1964. Bronze.*
2) Torso I, *1961. Bronze.*
3) Group of two figures B, *1964. Bronze.*

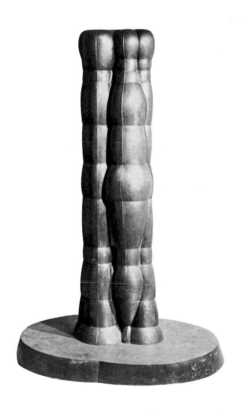

33

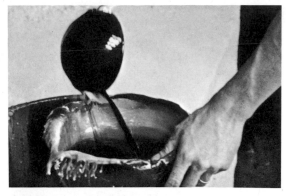

mould built, the founder must remove all traces of the seam left by the previous mould, correct surface faults and fit the wax with a core. The order of these tasks can vary. Small works are usually removed from the mould and rectified; a refractory core is then poured in. If inserted first this core could break on being removed from the mould with the wax because of its brittleness. Such an accident is unlikely to happen to the more flexible wax on its own. The core of a larger work however is poured before the work is removed from the mould, otherwise the pressure of the still liquid core against the inside of the wax skin might deform the wax model. Sometimes the wax is removed from the mould and immersed in water; this way the pressure exerted by the core is cancelled by the pressure of the water. It is important to remember that once the core is in place the statue becomes very heavy, and so is sometimes difficult to handle.

Retouching work on the wax is carried out with small needles and spatulas heated

The founder brushes a coat of wax into the gelatine mould.

either in a flame or electrically. Holes are filled with soft wax which is then reworked.

The core is either made from the same material as the investment used for the final ludo mould or in more porous clay. Long nails are fitted in to hold it in position once the wax has been melted out; they are embedded at the head in the ludo mould, then pierce through the wax into the core.

If the core is very thin it must be reinforced with lengths of wire, in a statue's legs for instance.

The air contained in the core must be able to expand or escape when the bronze is poured. If denied any other route the heated air will escape through the bronze forming small bubbles.

Once the core is in place, the wax, now perfectly finished, must be provided with a system of runners and risers. Bronze flows in through the runners, air and gases escape through the risers. Both are made from wax tubes which are firmly souldered to the wax model. A hole, called the *runner*

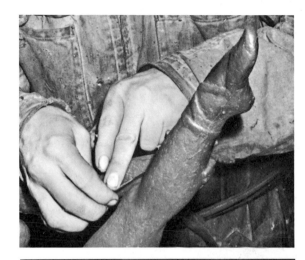

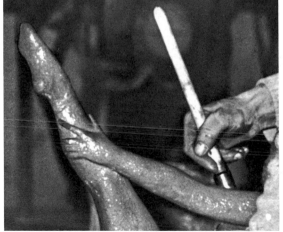

Retouching the wax.

Making runners and risers.

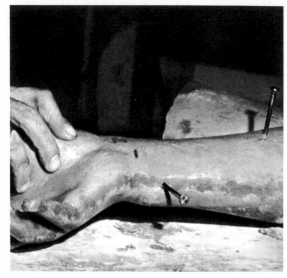

Putting in place the nails intended to hold the core in position.

cup or *pouring gate,* is left in the base of the work through which the molten metal is poured before spreading out through the runners. The system of runners is constructed in such a way that the bronze flows directly to the farthest corner of the work by means of the main runner. Side runners are attached perpendicularly to the main runner so that the metal fills the base first then rises stage by stage. The runners must be larger than the sections they are meant to fill. Care must be taken that these sections are not filled from top and bottom at the same time as this creates a risk of air traps. Risers are placed precisely at those spots where air is trapped inevitably;

The founder and his assistant put in place the system of runners and risers.

they usually come out on the sides of the mould towards the top. The risers have a much smaller cross-section than the runners. If the volume of air compressed is likely to be very small it is even possible to make do with a small lump of wax not connected to the exterior, but sufficient to provide a cavity for the air to expand in.

The ludo mould

The material used for the ludo mould and the core must be fireproof, suitable for application on wax, porous and strong enough to withstand the pressure of molten bronze. Clay has been used combined with various other materials, but nowadays

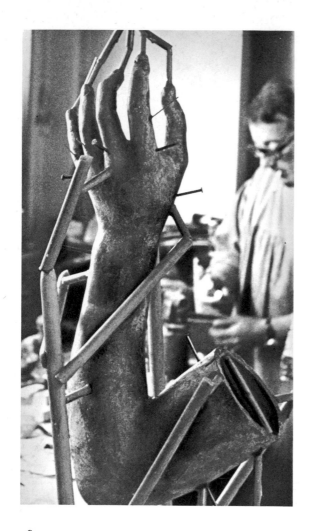

Equestrian figure with wax runners, risers and drains.

A wax ready for casting.

moulds are usually made from a mixture of plaster of Paris, brick powder and pulverized ceramic piping called *grog,* or other fireproof mixtures. Founders use powder produced by breaking up old ludo moulds to increase porosity and supply the necessary binding quality.

The first coat of investment is extremely important as the appearance of the bronze on its removal from the mould depends on it. It must by applied with a thick brush to a depth of about 1/4 in. (6 mm.), taking care to avoid air bubbles. As this layer is built up, each new coat must be applied before the previous one is completely dry to prevent flaking and to avoid cracks through which the bronze could seep during casting.

Traditionally the mould is then built from the base upwards, care being taken to leave the mouths of the complex of runners and risers near the top. In order to achieve a constant thickness the founder builds the mould in successive circular layers filling in every nook and cranny meticulously. The thickness of the mould depends on numerous factors such as the size of the work, the possibility of controlling the furnace temperature and so on . . .

For a life-size head, 1½ in. (38 mm.) thickness of mould must be allowed in

Horseman, *Benin bronze, second half on the 18th century.*

The first coat of investment mould is delicately applied to the wax with a brush.

addition to the first 1/4 in. when casting under good conditions, and up to 2 in. (51 mm.) if the foundry is poorly equipped. For larger works moulds can have a thickness of up to 5 in. (127 mm.), they must then be reinforced by strong wire netting.

The wax and its system of runners and risers . . .

When the mould is completed it must present an even surface free from holes and protuberances to prevent some parts from becoming crumbly through over-heating.

The furnace and its firing
The completed mould must now be baked until every trace of wax and moisture have disappeared. The smallest particle of wax

is completely covered with investment or ludo mould.

remaining in the mould will produce gases during casting and flaw the bronze.

The temperature necessary for melting wax can be reached in various ways, most commonly by using a furnace specially built for each casting.

The moulds are placed on two rows of bricks, pouring gates downwards. Small metal drains may be arranged between the bricks to collect the wax. A surrounding wall in brick cemented by a mixture of sand and clay is built round the moulds. The walls must be higher than the tallest mould. The top is sealed hermetically. Fires, usually of coal or oil, are lit between the bricks holding the moulds and the

The ludo mould.

The floor of the furnace.

walls. The moulds should not be touched directly by the fire; the heat is radiated from the walls. The firing up to 200 or 300° C should ideally be rapid. Some wax will drain from the moulds while these begin to dry out. When they are dry the temperature must be raised to 600° C; at this heat the remaining wax that has been absorbed by the moulds escapes from the pouring gate as scalding gas. This temperature must be maintained for several hours, or even days, depending on the size of the statue. Then the furnace is left to cool down slowly. In up-to-date foundries the temperature can be controlled throughout using a pyrometer.

Putting the armature in place . . . and completing the ludo mould.

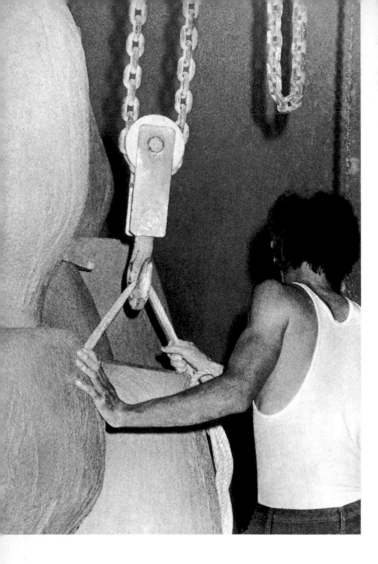

45

Once the wax has been melted out the empty moulds are surrounded with fireproof sand. Large moulds are buried individually.

For small pieces this operation can be carried out in a potter's kiln, or even by winding resistance wire round the surface of the mould. The mould is held with the pouring gate facing upwards and fitted at the base with a tube through which the wax drains. A sheet of asbestos is placed round the mould and the space in-between is filled with a powder of insulating material. The resistance wire is switched on and heats the mould, emptying it of wax. This process is not used in foundries but has been adopted by some artists who cast their own works.

Pouring the metal
The mould, now emptied of wax, is ready to receive the molten metal which will form the statue. But first it must be reinforced; so it is buried in damp sand to within about 12 in. (30 cm.) of its top. It is essential that the sand should not be too wet because dampness seeping through

Small moulds are buried side by side.

the walls of the mould could provoke an explosion or at least damage the statue. The founder will also take care to prevent grains of sand from penetrating the mould through the pouring gate.

When all the moulds are in position, the metal must be heated to melting point before it can be poured. For this there are two main types of furnace. In one sort the metal is melted in a *crucible* made of graphite and carborundum, a very tough agglomerate which withstands high temperatures. The furnace itself is composed of an exterior envelope of metal lined with a thick fireproof coating in the centre of which the crucible is placed, slightly raised from the floor so that the heat can spread evenly. Different fuels are used, but at present liquid gases and oil are most often chosen. The furnace must be heated to around $1000°$ C or more exactly $1100°$ C for bronze, 960 to $1010°$ C for silver and $1100°$ C likewise for gold. Of course the exact temperature depends on the precise composition of these different alloys but it is very important not to overheat. Overheating is quickly perceived as it has a visible effect on the colour of the metal.

The metal is not introduced until the crucible has become fiercely red-hot. Powder or other material to prevent the formation of gas is put in first, followed by small lumps of bronze which melt fast, finally larger lumps along with smaller fragments, this is so that the fusion, which is caused by heat radiated from the sides of the crucible, should take place with the minimum possible retention of air. If some lumps of metal are too big to fit in the crucible they can be heated on the top of the furnace as it is very easy to break up bronze above $500°$ C.

The founder can tell when a sufficiently high temperature has been reached by the colour of the metal, or by using a pyrometer, or even by dipping a ferrous metal rod into the crucible – if it is clean on removal the bronze is ready for casting. The deoxidizer has still to be added, usually in the form of phosphor-copper.

Large *tongs* with semicircular jaws are used to remove the crucible which is then lowered into a metal ring held by two

Pouring the bronze.

shafts one of which ends in a T, called a *shank and ring*.

When a foundry is equipped with a reversible furnace the metal is heated inside the furnace then poured into a steel container lined with an insulating coating which holds a crucible with extremely thin walls. The oxygen releasing substances, followed by the deoxidizers are then added, outside the furnace, and the crucible is carefully skimmed each time. When the surface of the crucible is perfectly free of cinders two men lift it up by the shank and ring; then, while the one holding the plain shaft maintains the correct height the other starts to pour the metal into the mould by turning the T-bar. This operation must be carried out very steadily without jerks or interruptions until the mould is full. A single jerk can jeopardize a whole casting irremediably by creating an air bubble which could be fatal enough to block the molten metal in one of the runners, thereby obstructing it beyond repair.

If these precautions are taken and if the runners have been nicely calculated the

The molten bronze is poured into the crucible.

metal will fill the space left by the wax perfectly.

When a very large quantity of bronze has to be poured into the same mould, the

50

Impurities which appear on the surface of the molten bronze . . . are removed by the founder.

operations described above still apply, the only difference being that the crucible is held by a system of pulleys; some crucibles can hold up to 500 kg. of metal. In the

old days, particularly during the Renaissance, large works were cast by the method still used today for casting bells, which Benvenuto Cellini has described very clearly in his *Memoirs*.

When the moulds are full they are covered over.

Foundry during casting. ▷

Opening the mould

Once the metal has solidified in the mould, the founder frees it using a chisel and mallet, removing one by one the armatures with which the mould was strengthened. Once he has reached the surface of the sculpture he usually completes the final stages by plunging the whole work in water. If the metal is still hot, a large part of the remaining plaster will crack open or at least soften. The statue can then be brushed clean, whereupon it will resemble the wax as it was just before receiving its first coating, that is to say with the whole system of runners and risers still in place. These are cut away with a pair of bolt cutters, or with saws which permit closer work without risk, leaving only small stubs to be filed down.

The nails are either removed with pincers or cut and driven inside. The statue can then be cleaned with a metal brush to remove some of the unattractive whitish-grey look which it had on leaving the mould.

Before starting retouching work it is usually necessary to get rid of the core which is inside the bronze. This is partly

Amida Buddha, *18th–19th century Japanese bronze.*

When cool the moulds are broken open.

to lighten the statue, but also to prevent the acid used to clean the bronze from being absorbed by the core, later to reappear on the surface as unattractive salts.

There is no difficulty in removing the core of a work which has a large opening in the base. Otherwise, small holes are pierced through which the core can drain,

having being made fluid by the introduction of water. As much of the core as possible, along with any metal armatures found in the cast, should be removed in this way.

Works which have been cast in separate sections are now welded together and the real finishing work can be undertaken,

The work appears, complete with runners and risers.

Remaining lumps of investment are removed with a chisel.

unless the sculptor decides that his work should stay as it is, which is sometimes the case.

Finishing

The first stage includes removing what is left of the runners and risers, filling the holes made by the nails and filing clean the surface of the sculpture. The remains of the runners and risers are removed with a *chisel* or by filing, the former is by far the preferable method as it can be used very precisely to retouch details without damaging the surrounding surface.

Holes are first enlarged – with an electric drill these days – and then tapped. A metal rod is driven into the hole and cut flush with the surface of the sculpture. This

Cleaned works ready for finishing.

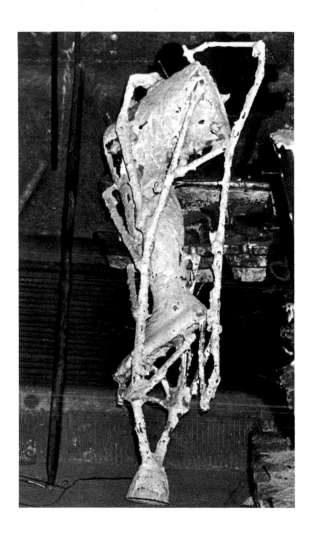

technique is valid for nail holes but equally for holes which have appeared for other reasons.

If as a result of accidents during casting large cavities have appeared, they are filled with a solder of an alloy similar to the metal of the sculpture itself. If a cavity is too big the missing part is recast in a mould taken from the statue, and then welded in place.

The runners and risers are sawn off.

The stubs are removed with a chisel.

Surfaces retouched in this way are then reworked by chisel or *burin* and dulled with matting tools in order to have a grain comparable to the rest of the statue.

When it is ready the bronze is thoroughly cleaned either with a fine sand blast or in an acid bath. A mixture of one part sulphuric acid to nine parts water usually suffices to give the statue a colour like brass, although it will still be rather blotched. A bronze can remain in this acid bath for several hours without deteriorating. The acid must then be eliminated by a thorough washing and neutralized with an alkaline solution, warm bicarbonate of soda for instance.

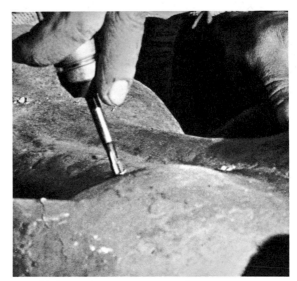

Small holes in the bronze are tapped.

Otto Freundlich, Composition, *1933, bronze.*

The patina

If a bronze is left in this state it will change colour gradually, becoming darker and finally almost black through oxidization. However sculptors usually prefer to control this colouring process, and above all to preserve a colour which pleases them. This is done by treating the bronzes chem-

The holes are filled with lengths of bronze rod . . .

. . . they are forced in, then sawn off.

ically, then polishing them, and occasionally even by applying a coat of lacquer.

The chemical reactions and thus the colours resulting from them are not yet fixed and permanent as they depend on a great many factors such as humidity, the action of the atmosphere, temperature, the quality of the bronze surface, etc. . . .

Here, however, are certain compositions in common use:
for a light brown bronze: liver of sulphur,
for a light green bronze: a solution of ammonia, sodium chloride, and sal ammoniac diluted in acetic acid,

 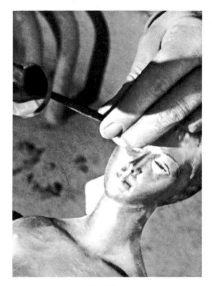 

The final stages of finishing:
matting — *chiselling* — *polishing*

for dark green bronze: a solution of nitric acid and copper nitrate diluted in water; sal ammoniac lightens the patina, copper chloride yellows it,
for antique green bronze: a solution of copper nitrate, hydrochloric acid and water.

The works are either immersed in the solution or have it applied with a brush, then, when the desired colour has been achieved, they are washed carefully. Finally the bronzes are given a coat of wax and polished.

Soldering bronze.

Constantin Brancusi, Cock greeting the sun, *1935, bronze.*

6 SAND-CASTING

This technique has certainly been in use since antiquity. At some periods it equalled or even surpassed the popularity of lost wax casting. Today it is used extensively in industrial casting. It is also greatly appreciated by artists who want to make several casts from the same original without having to tackle the many stages of lost wax casting. The piece to be cast, however, should be neither too large or too thick. As a general principle it can be said that relatively simple works intended to be cast in several copies, or works that need to be assembled with great precision afterwards, are most suitable for sand casting; while highly complicated relief work can best be dealt with using the lost wax process. As many of the steps are similar for both techniques, they will only be discussed briefly here.

The artist delivers his work to the founder. Here again, any one of several different materials may have been used but usually plaster coated with an insulating varnish is favoured. First of all the founder dissects the work, an operation rendered necessary by one of the biggest differences between the two methods of casting – the

Varnished plaster master cast prepared for sand casting.

Albert Rouiller, Composition, *1971, aluminium.*

number of runners and risers. Whilst in lost wax casting the structure of the runners can be extremely complex and stretch all the way round the statue to reach the farthest points, in sand casting it must be very simple and in one plane. Parts which are out of reach from this plane are detached from the work, re-arranged in the plane and connected to the central runner.

The different parts of the work are then spread out on the surface of one half of the mould making sure that the largest parts are placed so as to be reached by the

Pouring bronze into sand moulds.

The mould is filled with carefully sifted sand.

Half a mould with imprints.

metal first, i. e. at the farthest end from the pouring gate.

The bottom of the mould is filled with sand which is *tamped* (packed down) by hand, then vibrated by a pneumatic or electric machine giving it great homogeneity. The surface of the sand is dusted with talcum powder to prevent the master cast from sticking. The master is buried up

to its proposed seam, usually about half the work's depth. Then it is removed leaving a perfect imprint which will later be filled with bronze. To prevent damage on the removal of the master, areas with pronounced undercutting are moulded separately. The caster forms small blocks of sand to take imprints of these tricky surfaces which he later joins up to the main mould.

66

Delicate details are moulded separately . . .

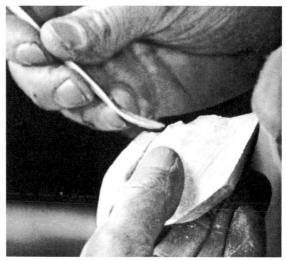

the fragment of tamped sand is sprinkled with talcum, retouched . . .

The main and side runners are cut into the mould. If the work is particularly large additional cavities are cut into the runner system to act as reservoirs for the molten metal, which, having filled the imprint of the master cast tends to shrink slightly; the molten metal held in reserve fills the space formed by the shrinkage. The system can contain more than one main runner,

it is nevertheless always constructed bearing in mind the rule that the metal must reach the farthest corners first. The risers are situated towards the outside of the mould and give onto its upper surface.

When the two halves of the mould are completed they are joined by an extremely precise mechanism and left in a drying-room to harden.

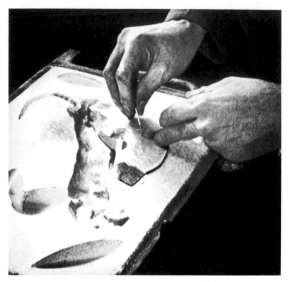

. . . and put in place on the other half of the mould.

First half of the mould with the master cast still in place.

Some works of large volume need to be fitted with a core. This is made by filling the female imprint, previously dried and dusted with separating powder, with sand which undergoes the same treatment as the mould, in which it is shut and packed hard. A layer equivalent to the eventual thickness of bronze is then pared off. The core is held in place by three metal sup-

ports. Also, a copper or steel tube pierced with small holes is fitted into the core for gases to escape through during pouring. All traces of moisture are eliminated by heating the drying-room to 450° C.

The moulds of small scale work are piled on top of each other to facilitate pouring the metal. The metal in its crucible is treated the same way as for the lost wax

Second half of the mould.

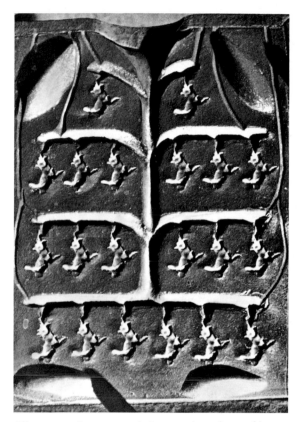

The system of runners and risers cut into the mould.

process before being poured into the pouring gates. Once again a steady uninterrupted pouring is indispensible for good results.

One knows the metal has solidified as soon as there is no longer a red glow from the pouring gate. The mould can now be opened and the sculpture removed.

It looks very different from a work retrieved from a lost wax mould; the surface,

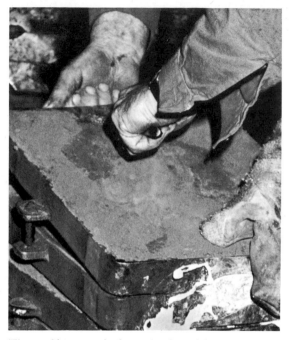

The mould is opened when it has been filled with bronze.

The work is removed from the sand.

when it is not covered with grains of sand, is shiny and in fact similar to its finished appearance.

The system of runners and risers is removed from the statue, the various parts welded together, and any traces of seam eliminated. The finishing and the patina are the same as for the lost wax process.

The removal of the statue deforms the imprint completely rendering the mould useless. So it is broken up and the sand retrieved to be used for new moulds.

70

7 VARIATIONS

Industrial and chilled casting

During the Bronze Age certain objects, like the heads of axes and spears, were cast in permanent moulds made of heat-resistant stone.

In our industrial era high-precision castings are mass-produced in steel moulds. The cost of making the moulds is nearly prohibitive. The combination of price, technical difficulty and a certain prejudice agains industrial methods has probably prevented the process from being adopted until recently. Also, it is a method which produces solid castings, necessarily limited in size.

Today, however, in this second half of the 20th century when a trend towards popularizing art, in the best sense of the word, is becoming clearly discernable, there is at least one case of a sculptor-founder using injection moulding to produce mass-produced high-precision works for wide diffusion. This is Miguel Berrocal and his "assembled multiples".

Having sketched a statue, Berrocal breaks it down into a number of parts to form the first prototype. The adjustments of various elements are studied with care, following which drawings are produced and executed in metal; this is the final prototype. When it has received its finishing touches a female imprint is taken of each component part. For this, ultra-sounds are used or alternatively electrodes, guided by a pantograph, which hollow out the metal reproducing the original in all its details. This operation is carried out in an oil bath so as to permit the use of shrink-proof hardened steel. It takes three separate electrodes working in succession to complete the job.

The basic difference between chilled casting and injection moulding lies in the way the metal is introduced into the mould. In the former the mould is fed by gravity, so that the metal flows from the top downwards. In injection moulding, on the other hand, it is possible to introduce the molten metal from any side, thereby eliminating many problems due to distortion and shrinkage.

The machine tools used for injection moulding are either fully automatic, in which case the metal flows from the crucible where it was melted to the mould without human intervention, or semi-automatic, the metal then being poured by hand into a piston tube which compresses it and injects it forcibly into the mould.

Only with a process such as this, producing remarkably regular and well finished castings, is it possible to envisage diffusing a thousand or even more editions of a multiple statuette.

The only snag with it, apart from its expense, is that neither the nobler metals nor steel can be cast this way.

For these the artist must resort to a perfected version of the lost wax process, using this time a mould made of rubber.

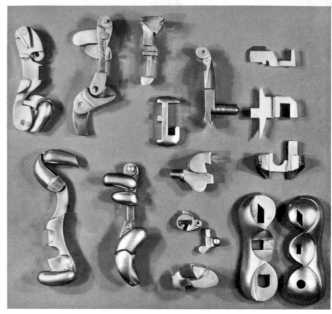

The sixteen elements of Berrocal's Romeo and Juliet *cast by injection process.*

Unlike gelatine, rubber is deteriorated but little by heat; it is decomposed by polymerization. To obtain results as perfect as by injection moulding, it is best to have the molten metal sucked in by a vacuum pump.

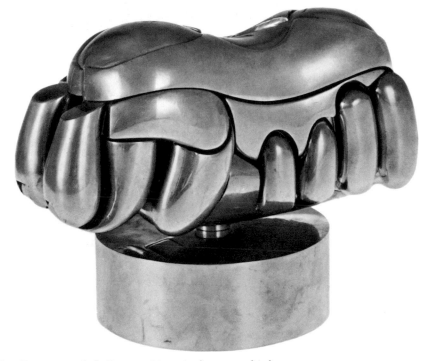

Miguel Berrocal – Romeo and Juliet, *1966–1967, bronze multiple.*

After polishing the various parts are ready for assembly.

Casting with expanded polystyrene

Modern materials permit new casting techniques which are still rarely used but open up very exciting possibilities. Expanded polystyrene calls for a type of casting which resembles both the lost wax process and sand casting, and which, on a different scale, may remind you of the technique for casting very small objects made in solid wax by the lost wax process.

The artist fashions his work in expanded polystyrene, a white material which is extremely light and melts very easily. He works with a heated resistance wire with which he can cut, trim, pierce and in other words sculpt this extraordinary material just as he pleases. Although up till now there has been no limit to the shape of works made this way, they are usually made solid, and thus slender, to avoid distortion. However it is quite possible to fit a core.

When the original is finished it is placed in a mould of specially composed sand which sets very quickly, becoming as hard as cement.

Once the mould is ready, bronze or another alloy, aluminium for instance, is poured directly onto the expanded polystyrene which it melts away while immediately taking its place.

The mould is opened and the work removed as for an ordinary sand mould.

Heracles fighting. *Greek bronze.*

8 THE GREAT AGES OF ART FOUNDRY

Antiquity and the Middle Ages
The history of bronze from its origins to the present day can only be written by retracing some seven thousand years of civilization. Ever since the discovery of metal, casting techniques have spread across the world and museums today are veritable bronze warehouses stored with work from all periods of history.

The uses of bronze have been so many and varied that they cannot all be covered by the title "Art Foundry" alone. Can indeed the marvellous buckles found in Burgundian tombs be considered as art?

And what about the ornaments with which 17th and 18th century cabinet-makers adorned their work? And in what category are richly decorated cannons to be placed?

Not even an encyclopedia could encompass the subject, so we shall concentrate here on the "golden ages" of bronze and deal but briefly with the minor periods.

The first oriental bronzes were cast mostly in India and China, notably the bloated dragons and fat warty toads which are still made in more or less the same mould and are fine examples of the extraordinary skill that Chinese artists acquired in working metal and other materials.

In the province of Luristan, to the west of Iran, and in the north of the country, lay concealed a complete fauna of bronzes. Although cast between the 11th and 6th centuries B.C., these works were only revealed to European eyes during the second quarter of the 20th century. The oldest among them, small figures of men and animals, are in a style which is both realistic and imaginative. They were mostly

Antonio Pollajuolo, Grammar, *detail from the tomb of Sixtus IV. Bronze relief.*

cast by the lost wax process and nearly always without a core. The animals they represent: stags with long antlers, bisons and ibexes, belong, in fact, to the fauna of the region, which seems to suggest that this art had purely local origins, at least until the arrival of Alexander the Great.

Egypt, which reached a very high degree of civilization long before our era, as is attested by magnificent remains, attached great importance to the arts. It encouraged the development of a remarkable tradition of bronze statuary. The most beautiful of the works that have come down to us were cast during the Saite period (665–525 B.C.); they include the seated figure, or Horus, now in the Louvre.

The earliest works cast in bronze were found in Crete and date from the first half of the second millenary before Christ. They are daggers, discovered in Mycenean

Cat, Egyptian, bronze.

Lillebonne Apollo. Greek bronze. p. 77.
Cnidian Aphrodite. Greek bronze.
Apollo of Piombino. Greek bronze.
Donatello, David. *Bronze.*

tombs, with cast bronze blades hammered and decorated with inlaid gold, silver and niello. This inlay technique was originally Egyptian, but it was widely used in Greek statuary for a long time.

Mycenean and Cretan bronze statuettes never equalled the perfection or design precision of the dagger blades. This is due in part to the fact that they were cast in solid bronze but also, and mostly, to the influence of another art, terracotta. In fact the bronzes look as if they were made in terracotta. They were not retouched and so bear the marks of modelling and occasional traces of mould seam.

The pre-Hellenistic and high Archean periods produced principally everyday objects: weapons, cups, plates, mirrors, lamps and amphorae. However, the oldest known example of hollow casting, known as the "Karlsruhe female head", dates from the second half of the 7th century B.C.

As time went by, styles peculiar to the various centres began to assert themselves

Horned god (Apollo, Alasiotas), Cyprus. Solid bronze.

—in Sparta, Arcadia, Aegina, Athens, Samos and also in Magna Graecia, at Locri and Tarentum. The style of work from the last two cities lags clearly behind that of the metropolis, and its quality is inferior.

The bulk of the production of the various studios of Ancient Greece was made up of small bronzes. Very likely the great popularity of marble statuary delayed the first attempts at large scale bronze casting.

Male and female caryatides holding mirrors and dishes form the principal, and perhaps the most attractive production of the art of the founders of the period, apart from a few larger works found mainly in Athens.

From 450 B.C. onwards bronze statuettes become surprisingly rare. Museum collections bear witness to this. The new fashion for large scale bronzes through which the great masters Phidias, Polycleitus, Pythagoras and others are known must have hastened their disappearance. The influence of these sculptors gradually overshadowed that of the regional workshops.

A new image of man then appears—the stadium athlete. He is portrayed in full action and not reconstructed from an abstract formula like the precise earlier works which sometimes strike us as unattractively dry. Lysippus with his Apoxyomenus of Ephesus and Praxiteles with his nude Aphrodite illustrate this new realism magnificently.

The conquests of Alexander the Great, while helping spread Hellenistic culture across the world, also brought about the creation of new centres of art in Egypt and Syria. Yet the classical tradition, as exemplified by the masterpieces of the great sculptors, continued to impose stylistic unity on Greek art.

Small scale works now consisted mainly of copies of the larger statues that kept appearing in a variety of new shapes and were usually inspired by Godlike figures such as Alexander, the poets, philosophers and orators.

Another minor art-form was developing at the same time that was to become widespread: namely representations of popular

Vix vase (fragment), Magna Graecia. Bronze.

Head of the emperor Constantius II, *late Roman. Bronze.*

Chinese art. Tang period 618–907. Bronze.

or grotesque characters—wandering merchants, priests, beggars, etc.

This abundant production continued way beyond the Hellenistic period, without break, in fact, up to Roman days. The passage from one to the other is imperceptible.

82

Roman sculptors were satisfied with copying Greek originals. They went about it conscientiously but without feeling; their copies are clumsy and heavy. For collectors they manufactured endless reproductions of masterpieces from the great Athenian period, or cast in bronze the features of this or that tribune or general. Perfect likeness was the object, so that these works were often remarkably uninspired.

The drive for higher productivity pushed the Roman bronze foundry-man into certain artistic malpractices which were later to reappear during the 18th and 19th centuries. They went so far as to cast standard bodies for interchangeable heads, the latter commissioned by the client!

Throughout the whole of the Middle Ages bronze statuary stagnated, with the exception of certain masterpieces like the doors of the San Zeno basilica in Verona, and those of the Gnieznienskie in Poland.

Early Christian metal-work, The toilet of Venus, *late 4th century. Cast silver.*

During this period a most impressive era of bronze work – which aroused the admiration of Europe – flourished in Africa, in the Benin kingdom. Edo, now Benin City in Nigeria, was then the capital of a kingdom which spread over a large part of the Guinean coastline to the west of the Niger.

Benin art, like the art of Ife, its immediate precursor, radiates a sense of peace and calm, in marked contrast with other African art-forms. The earliest known bronzes date back to the 12th and 13th centuries. They are heads of kings, very finely cast by the lost wax process. Small holes on the cheeks and surrounding the forehead, and lips, once held real hair and beards which have since disappeared. Later the castings became more varied. Cast bronze reliefs appeared next to the many in the round figures of warriors, horsemen, royal messengers, etc. These bronze panels decorated the porch pillars of the royal palace which Booman in his *"Voyage"*, published in 1704, describes thus:

"We must not forget the king's court which takes up the largest part of the village. We enter first a very long gallery, supported by fifty-eight planks, not pillars, about twelve foot tall. At the end of the gallery is an earth wall with three doors, one in the middle and one at each extremity; above the middle door is a wooden tower shaped like a chimney-stack, which is sixty or seventy feet high. At the top of this tower a bronze serpent hangs suspended, head down. This serpent, being so well cast and such a simple representation of a living serpent, is the most precious thing I have seen in Benin. On entering the door we find ourselves in a large place, about a quarter of a league square, enclosed by walls of earth, but not very high. Leaving this place we find another gallery similar to the first but without walls or tower. There are but two doors, then yet a third gallery different from the others in that where before there were planks there are now statues; our guides point out merchants, soldiers, stokers etc. Beyond a white carpet we see eleven heads of men in copper, more or less of the same mould; on each is an elephant's tooth. These are a few of the king's idols."

Female images are very rare in the art of Benin. Their attitudes and physical

Adam and Eve driven out of Paradise, *fragment from a bronze door at the San Zeno basilica, Verona, late 11th–late 12the century.*   *p.85*

Lorenzo Ghiberti. Moses receiving the tables of the law *(detail), 1425–1452. Gilt bronze.*

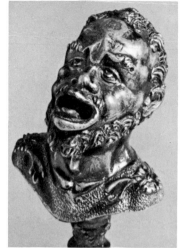

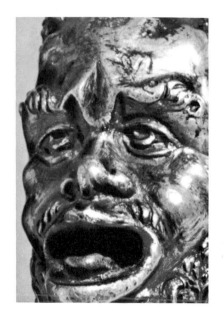

Riccio, Kneeling satyr, *late 15th century. Bronze.*

Riccio, Oil lamp, *late 15th century. Bronze.*

Riccio, Oil lamp *(detail).*

87

Benvenuto Cellini, Bust of Cosimo I *(detail), circa 1535. Bronze.*

type, which sometimes resemble works from Asia Minor, cause one to wonder about the communications that existed between civilizations in those days; no answer has yet been found to such speculation.

The Renaissance in Italy

1400 marks the beginning of a rebirth of the art of bronze in Italy. That year Florentine merchants organized a competition for the decoration of the second portal of the baptistry; Lorenzo Ghiberti won the commission. Sixteen years later the major guilds were authorized by the town-councillors to erect statues in bronze instead of stone in the guild-hall of Or San Michele. For more than half a century commissions flowed into Ghiberti's atelier where most of the young artists of Florence received their training. Donatello worked there for a few years, but soon split with the master who was too bound up by the traditions of his former craft of silversmith and too meticulous for Donatello, the creator of David.

One of the atelier's last pupils was Antonio Pollajuolo to whom we owe nota-

Christ entering Jerusalem, *buckle from a Burgundian swordbelt, 5th or 6th century. Bronze.*

bly the "Hercules and Antaeus". This little bronze illustrates perfectly the possibilities of small scale work for violent expression by gesture and shape.

One of Donatello's pupils, Bertoldo di Giovanni, seems to have been the first Renaissance artist to specialize in small scale bronzes, reliefs and medallions.

No doubt there existed furnishers of a less exalted style of bronzes, if we take the hint from this verse sung by master-sculptors at carnival time:

He desirous of the joys
Of our charming statuettes
Whether pedestal employs
Or by bed the statue sets,
Figurines formed by our art
Everywhere rejoice the heart.

Given the context of a carnival, it is easy to imagine the sort of subject depicted...

Donatello was to apply his overflowing energy to the making of monumental sculpture. He travelled the length and breadth of Italy during the whole of the first half of the 15th century and up till his death in 1466, creating workshops in Rome, Padua and Sienna, and training disciples all over the country. Although worthy, the works of these pupils never equalled the power of their master's forceful compositions, built on the monumental scale he favoured. Donatello rarely cast his own work; Cellini was later happy to point out that the David's imperfections were due to poor casting technique.

The foundries of Antonio Pollajuolo and Verrocchio were rivals at that time. Their artist clients supervized the work most carefully and insisted on being present in the foundry when the bronzes were finished. This explains the inscription on "Bellerophon" by Bertoldo di Giovanni: "Bertoldo modelled me, Hadrian cast me."

Bertoldo di Giovanni and another pupil of Donatello, Bartolomeo Bellano, then worked together on the bas-reliefs at St. Anthony's basilica, Padua. The style of Bellano's work is rather heavy and labored. His own pupil, Andrea Briasco, called Riccio, surpassed him easily with pieces like his paschal chandelier, a huge work nearly four meters high decorated with bas-reliefs depicting a variety of scenes featuring satyrs and cupids, as well as with small bronze ornaments of individual figures taken from the chandelier. Numerous copies of these were cast, the most popular models being undoubtedly manufactured by several Padua workshops. It is thus extremely difficult to authenticate them with any certainty.

After Donatello's death equestrian statues like the Gattamelata of Padua caught the fancy of the Italian rulers. Pollajuolo sketched a project for Lodovico il Moro in which Francesco Sforza can be seen in armour, mounting a horse rearing up above a vanquished enemy. His sketch was taken up again by Leonardo da Vinci and even modelled, but it was never cast.

Verrocchio undertook the Colleoni monument which he handed over to the founder Leopardi.

Parallel to these monumental projects there was a flourishing trade in small bronzes as already mentioned. In addition to the centres of Florence and Padua, mention should be made of Mantua, where Piero Jacopo Alavi Bonacolsi, known as l'Antico, worked; he was a fervent admirer of Greek statuary which he imitated time

Giovanni Bologna, Horse at the piaffer, *second half 16th century. Bronze.*

91

Antonio Susini, Wild boar, *second half 16th century. Chiselled bronze.*

Augsburg school, Lion, *second half 16th century. Chiselled bronze.*

and again with miniature Apollos, Venuses and Hercules.

Antonio Lombardo's soft round forms foreshadowed 16th century Venetian painting.

Unrivalled, the personality of Michelangelo Buonarroti dominated the 16th century, and the sack of Rome in 1527 only contributed forcibly to his influence by dispersing throughout Italy scores of

artists whose styles had been formed at his contact. Jacopo Tatti, called Sansovino, was one of them. He established himself in Venice where he became head-foreman of the St. Mark's foundry. The bas-reliefs narrating the story of St. Mark were created there, as were the sacristy doors and the four statues of seated evangelists on the balustrade of the St. Mark's choir.

The Italian Bronze School was now expanding fast and even if great monuments were not ordered every week small statues were polished off daily by artists like Tiziano Minio, Alessandro Vittoria and Gerolamo Campagna, who, at the same time, were not ashamed to undertake more common objects such as firedogs, salt-cellars, door-knockers etc. . . .

In Venice, Niccolo Roccatagliata, a friend of Tintoretto – he made wax maquettes for him – specialized in "putti". These were stylized figures of heavy-lidded round-cheeked little boys of indeterminate age, their hair gathered at the forehead and temples. They were fashionable for a long while, to such an extent that the expression "a Roccatagliata putto" came to mean the style and not necessarily a work by the sculptor himself.

Florence, during this first part of the 16th century, did not participate in the general craze for bronze statuettes, perhaps because of Savonarola's ringing condemnation of this kind of frivolity.

But in 1537 Cosimo I ushered in a new era which turned Florence into the capital of the arts during the second half of the

Bernini, Pope Urban VIII, *1642–1647. Bronze.*

93

16th century. In his *Memoirs* Benvenuto Cellini paints a lively picture of an artist's life at the court of Cosimo. Apparently, things were not always rosy; financial difficulties kept cropping up despite the princes' interest in their artists. Nor did any entente cordiale reign between the ruler's protégés. Perhaps Cosimo hoped this way to stimulate them to even greater efforts to obtain his favours.

Cellini's *Memoirs* abound in informative details. Of special interest to us are his descriptions of casting. We learn that with rare exceptions sculptors had their work cast by bell-founders. This seems to have lead fairly often to unwelcome misadventures, because the craftsmen knew nothing of the lost wax process or of the specific problems of statue casting. Benvenuto himself occasionally had to rely on bell-founders, either in Paris or in Florence where the master-caster Zanobi reigned supreme. This renowned founder once cast successfully a reduced version of the Medusa from the foot of the famous

Gaspard Marsy, Boreas carrying off Orithya, *18th century. Bronze.*

Perseus. Cellini's description of the casting of the statue itself is amusingly vivid and very interesting:

"Thus, having recovered my vigour of mind, I exerted all my strength of body and of purse, though indeed I had but little money left, and began to purchase several loads of pine-wood from the pine-grove of the Serristori, hard by Monte Lupo; and whilst I was waiting for it, I covered my Perseus with the earth which I had prepared several months beforehand, that it might have its proper seasoning. After I had made its coat of earth, covered it well, and bound it properly with irons, I began by means of a slow fire to draw off the wax, which melted away by many vent-holes; for the more of these are made, the better the moulds are filled: and when I had entirely stripped off the wax, I made a sort of fence round my Perseus, that is, round the mould above-mentioned, of bricks, piling them one upon another, and leaving several vacuities for the fire to exhale at. I next began gradually to put

Jean-Antoine Houdon, Diana, *1790. Bronze.*

Honoré Daumier, Ratapoil, *circa 1890. Bronze.*

on the wood, and kept a constant fire for two days and two nights, till, the wax being quite off, and the mould well baked, I began to dig a hole to bury my mould in, and observed all those fine methods of proceeding that are prescribed by our art. When I had completely dug my hole, I took my mould, and by means of levers and strong cables directed it with care, and suspended it a cubit above the level of the furnace, so that it hung exactly in the middle of the hole. I then let it gently down to the very bottom of the furnace, and placed it with all the care and exactness I possibly could. After I had finished this part of my task, I began to make a covering of the very earth I had taken off, and in proportion as I raised the earth, I made vents for it, which are a sort of tubes of baked earth, generally used for conduits, and other things of a similar nature. As soon as I saw that I had placed it properly, and that this manner of covering it, by putting on these small tubes in their proper places, was likely to answer, as also that

Ife, Portrait of a king or Oni, *12th or 13th century. Thin lost wax bronze.*

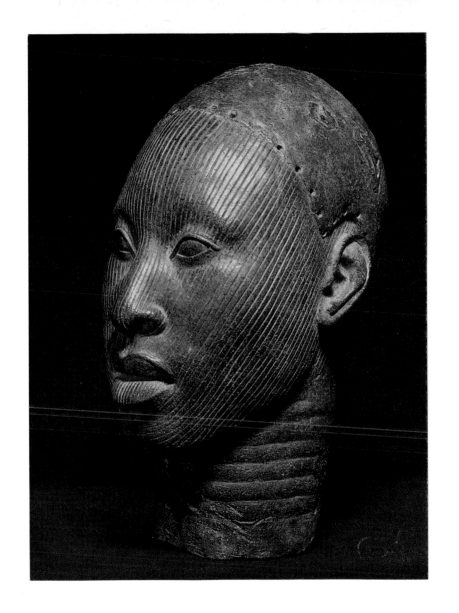

97

Auguste Rodin, Man walking, *1911. Bronze.*

my journeymen thoroughly understood my plan, which was very different from that of all other masters, and I was sure that I could depend upon them, I turned my thoughts to my furnace. I had caused it to be filled with several pieces of brass and bronze, and heaped them upon one another in the manner taught us by our art, taking particular care to leave a passage for the flames, that the metal might the sooner assume its colour and dissolve into a fluid. Thus, I with great alacrity, excited my men to lay on the pine wood..."

At this juncture Cellini had to take to his bed because of an illness. During his absence the founders he had hired ignored his instructions, finally dragging him from his bed declaring that the casting had failed completely.

"I went directly to examine the furnace, and saw all the metal in it concreted. I thereupon ordered two of the helpers to step over the way for a load of young oak, which had been above a year drying. Upon his bringing me the first bundles of it, I began to fill the grate. This sort of oak makes a brisker fire than any other wood whatever; but the wood of elder-trees and pine-trees is used in casting artillery, because it makes a mild and gentle fire. As soon as the concreted metal felt the power of this violent fire, it began to brighten and glitter. In another quarter I made them hurry the tubes with all possible expedition. Then I caused a mass of pewter weighing about sixty pounds to be thrown upon the metal in the furnace, which with the other helps, as the brisk wood fire, and stirring it sometimes with iron, and sometimes with long poles, soon became completely dissolved. Finding that, contrary to the opinion of my ignorant assistants, I had effected what seemed as difficult as to raise the dead, I recovered my vigour to such a degree, that I no longer perceived whether I had any fever, nor had I the least apprehension of death. Suddenly a loud noise was heard, and a glittering of fire flashed before our eyes, as if it had been the darting of a thunderbolt. Upon the appearance of this extraordinary phenomenon, terror seized on all present, and on none more than myself. This tremendous noise being over, we began to stare at each other, and perceived that the cover of the furnace had burst and

Jacques Lipchitz, Young girl with tress. *Bronze.*

flown off, so that the bronze began to run. I immediately caused the mouths of my mould to be opened; but finding that the metal did not run with its usual velocity, and apprehending that the cause of it was that the fusibility of the metal was injured by the violence of the fire, I ordered all my dishes and porringers, which were in number about two hundred, to be placed one by one before my tubes, and part of them to be thrown into the furnace; upon which all present perceived that my bronze was completely dissolved, and that my mould was filling; they now with joy and alacrity assisted and obeyed me."

Next to Giovanni Bandini, Vincenzo de Rossi and Vincenzo Danti, who all participated in the decoration of Francesco de' Medici's study, the court of the Medici also welcomed during the second half of the 16th century a Flemish artist, Giovanni Bologna, sometimes known as Giambologna, whose reputation spread like wildfire. Thanks to his brilliant exploitation of the potential of small scale bronzes his work was diffused throughout the courts of Europe. His most famous work "Mercury" is in the Bargello. In both figure and group sculpture this Flemish artist, who was greatly influenced by Michelangelo, achieved a well-balanced harmony. He was more interested in "Beauty" than in modelling the likeness of this or that particular hero. As he was not himself a founder, however, his massive output necessitated the creation of an organization almost on an industrial scale, where his own part was limited to making the models in wax or clay, for small works at least, leaving the perfecting of his statuettes to the care of casters, founders and metal-workers.

The master-founder Antonio Susini worked for Giovanni Bologna alone, and made a fortune doing so. He perfected a commercially ingenious process which made it possible to cast a horse, a horseman and the horseman's head separately. These elements could be assembled in many different combinations and only the man's head would normally be made to order. These statuettes soon became the fashionable gift for nobles to offer their friends,

Constantin Brancusi, Mademoiselle Pogani, *1933. Bronze.*

Alexander Archipenko, Head, *1913. Bronze.*

and an essential "souvenir" for strangers to bring home from Florence. Equestrian models were not the only items to be prized thus, genre scenes representing everyday characters like bird-catchers and bagpipe players also soon found buyers.

This proliferation, and the growing estrangement between the artist and his cast work, were probably responsible for the decline in the quality of work in bronze, an evolution which was to be consummated in the 19th century by the introduction of mass-production techniques.

The Renaissance in Germany

Some noteworthy bronzes were cast in Germany during the 16th century, yet bronze was never as popular there as in Italy. Traditionally German sculptors worked wood and stone; their only contact with bronze foundry-men would be on the rare occasion of a wood original being ordered for casting. Nevertheless, the Vischer atelier at Nuremberg was very active. Amongst other works it was responsible for the magnificent gothic reliquary at St. Sebald's which was commenced in

1488 and completed in 1519. It also produced a great many statuettes most of which have a hard look about them that betrays their wood origins.

Two families and two towns attained a certain mesure of renown for bronze statuary in 16th century Germany: the Reisingers at Augsbourg and the Labenwolfs at Nuremberg. The former were influenced by the Italian art which had been introduced to Germany by their masters, the Fuggers. They cast large richly decorated fountains, sometimes using a technique which was very popular in Italy at the time – casting from nature. This method made it possible to catch the plastic beauty of small animals, reptiles, birds, insects and even drapery by taking an imprint directly from the beast or the object itself, instead of first fashioning a reproduction in clay.

The best known piece by the Labenwolfs is the famous "Cupid" of the Town-hall fountain in Nuremberg.

At Mühlau, near Innsbruck, the Archduke of Austria, Maximilian I, created a foundry from scratch, with the sole purpose of having the statues he intended for

Raymond Duchamp Villon, Horse, *1914. Bronze.*

Antoine Pevsner, Spiral construction, Le lys noir, *1943. Bronze.*

104

his tomb cast. They were to include all his ancestors from the House of Hapsburg and also several hundred statuettes of saints. The task was shared out between three craftsmen; Georg Ködelere designed the works which were modelled by Leonhardt Magt and cast by Stefan Gold. Despite this team-work the task was never completed. Only 23 works were cast. The vigour of their modelling, which owes a lot to the chiselers added touch, and the quality of the bronze, lend them great majesty.

Modern times

Colossal statues were erected all over France in the 17th and 18th centuries; most of them have since disappeared; they were melted down, often to make cannon, when the personnages they portrayed fell from favour. Such was the fate notably of the eight meter high equestrian statue of Louis XIV, modelled by François Girardon, who decorated the Versailles gardens with sumptuous compositions, and cast in one go by the Paris founder Balthazar Keller. A few years later Edmé Bouchardon was equally unlucky with an equestrian statue of Louis XV which had been completed by Jean-Baptiste Pigalle. The Revolution did away with them both. These monuments were often copied on a small scale, either by the artist himself, or by his disciples, friends or even rivals. These small scale copies, which also sometimes included imitations of antique work, were carried out without mechanical aids and were rarely faithful to the original. Their quality depended on the skill of each of the three craftsmen involved: the modeller, the founder and the chiseler. To add to their attraction they were often in whole or in part gilded, especially during the 18th century.

Gilt bronze then came into its own, particularly in decorative art, as an ornament for furniture or utilitarian objects. French craftsmen, organized in two guilds, the founder-chasers and the chiseler-engravers, collaborated with cabinet-makers like André-Charles Boulle, Jean-Henri Risener and C. Beneman, who placed orders with them for all kinds of furniture decoration. Not infrequently they also manufactured fire-dogs, chandeliers and clocks which were exported throughout Europe.

Jean Arp, Ombre chinoise, *1947.*

Aristide Maillol, Woman getting out of bath, *circa 1920. Bronze.*

These pieces usually went unsigned; they were the result of team-work by several specialists. The many different payments made in 1766 for a pair of fire-dogs surmounted by lions which had been ordered for the palace at Versailles illustrate clearly the division of labour: Bureau, Pigalle and Boureffe supplied projects in wax and wood, Boizot modelled the couched lion in clay, Michaud made a plaster cast of the lion, Martin was responsible for the clay and wood version of the decorations of the base. Forestier cast the various pieces in metal, which were chased by Thomire, Coutelle and Boivin. Lastly the work was gilded by Galle.

Even the greatest among the artists of the time did not scorn this minor form of art; Houdon, for instance, modelled a pair of chandeliers figuring four cupids bearing a cannon. Active throughout the second half of the 18th century and well into the 19th, he cast, or had cast, a great number of portraits in bronze, of aristocrats primarily, but also of politicians and figures from the worlds of art and literature. Houdon was both sculptor and founder, and even if some of his work was destroyed by the blind fury of the Restauration – the monumental bronze of Napoleon I for instance – what remains constitutes a lively portrait gallery of his time in both bronze and marble. That he was interested in casting and allowed the technique to play an important role in his life is evident from the following memoir which he sent to a friend on 20 Vendémiaire (11.x.1794):

"...Summing up my life's work, I see that I have devoted myself to only two subjects – anatomy and statue casting – which have taken up everything I have earned and indeed my whole life, which would better have served my country had I been backed or in possession of a private fortune. Lodging for a long while at the municipal workshops, I made the most of the situation by becoming both sculptor and founder (today these two professions are always exercised by different people), reviving in my country this useful art which was in danger of being lost as the old founders were all dead when I took over; I built furnaces, I trained workmen, and after many fruitless and expensive attempts I succeeded in casting two statues myself: the Diana, of which I still possess a copy, and my Frileuse. Chased from these studios by Breteuil in 1787, within three weeks I had bought the house opposite, built new furnaces and cast my Apollo... I can be regarded as both sculptor and founder: as the former I

Aristide Maillol, Final study for Cézanne monument, *1925. Bronze.*

create; as the latter I carry out a durable version of another's creation, and this at less expense than anyone else – since, having had nothing but my own savings to invest in the art, I have inevitably learned to reduce or cut out unnecessary expense."

Elsewhere Houdon mentions "a founder of great merit"; this must certainly have been the chaser and founder Philippe Thomire, whose bronzes, mostly statuette reproductions of works by contemporary masters, were beautifully chased and coloured.

At the beginning of the 19th century two techniques appeared which were to revolutionize the art of small scale bronze statuary, and eventually ruin it too. The first was the massive utilization of sand casting, which almost caused the lost wax process to disappear.

The second was the invention in 1839, by Collas, of a mechanical reducing apparatus which was capable of reproducing a statue with great accuracy on any scale. One can imagine how easy it became from then on to churn out copies of all subjects. The production of small scale bronzes then became practically industrialized, a

Henry Moore, Reclining figure, *1951. Gilt bronze.*

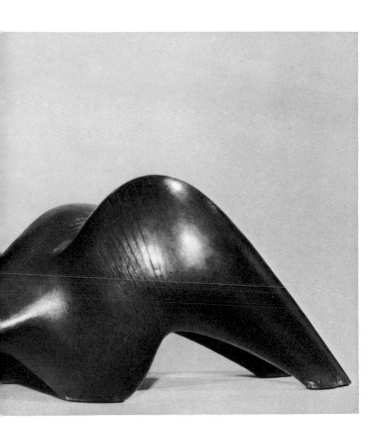

fact borne out by the reversal in status of sculptor and founder. It was the latter who now held the artist under contract and a bronze edited by a founder like Barbedienne was more sought after than a work by Barre, Pradier or even Barye. Because an edition of such statuettes was unlimited, the bronzes were not numbered. As a result their quality and market value soon fell. In fact poorly finished castings sometimes made of cheap alloy or even of another material with just a coat of bronze paint now flooded the market.

Antoine-Louis Barye alone still had faith in the lost wax process as a means to give life to his animal sculptures. He either cast his bronzes himself or entrusted them to Honoré Gonon, an idealistic and talented founder.

Monumental sculpture was still going strong at the beginning of the 19th century. Artists like François Rude, James Pradier and Philippe Lemaire sometimes laid aside stone and marble to attack works in bronze which were taken over by the founders of the period, Richard, Eck and Durand, for instance, who had a monopoly over production in Paris for a time.

But at the end of 19th century three artists were to transfuse new blood into the art of bronze: Daumier, Degas, and most important of the three, Rodin. The first two were painters, but modelled figurines at intervals throughout their careers. These figurines were in fact cast after the artists' deaths and their modelling brilliantly rendered by the founder Hébrard who reproduced the aspect of the original material with remarkable precision. This is also true of Rodin's work in which the sensuous quality of the original clay surface and even the sculptor's fingerprints appear in the bronze. Rodin went over his large waxes carefully, and his founders, Alexis Rudier in particular, worked hard at conserving exactly the aspect the master imparted to them. Everything worth writing about Rodin has already been written, there is no point in analysing here the influence he exerted on contemporary sculpture, or the reasons for this influence. Suffice it to say that he was engaged in the same quest for Truth and the origin of things which has characterized so many 20th century artists.

Nearly all the leading contemporary painters have tried their hand at three-dimensional work, occasionally launching new trends in modern sculpture. Many of them have worked in metal, restoring the lost wax process to its former eminence. In France, Germany, and Switzerland too, big foundries have devoted themselves to art work, in the wider sense of the term, to the exclusion of everything resembling industrial casting, whether of cannon or bells. While these foundries have of course modernized part of their equipment, replacing coke, say, by oil or electricity, they are nevertheless faithful to the traditional techniques of casting.

This continuity is worth stressing, for it is one of the essential characteristics of art foundry. It is in the finishing that the biggest changes can be observed. We have already mentioned how faithfully Rodin's work, or Bourdelle's for that matter, was rendered in bronze, how the aspect of the

Julio Gonzales, Small sickle, *1937–1938. Bronze.*

Barbara Hepworth, Single form, Aloe, *1969. Bronze.*

III

Henri Presset, Haut-Relief à l'œuf, *1967. Bronze.*

clay was reproduced in the more durable material by skilful use of the file. At the other extreme there are works by Jean Arp, Barbara Hepworth and even Constantin Brancusi, which are so highly polished that the metal reflects even the slightest play of light. Alberto Giacometti, Germaine Richier and others prefer the rough unfinished aspect of a bronze fresh from the mould. Yet others, Laurens, Archipenko and Miró, for example, apply a polychrome finish to their works under which the aspect of the original material disappears.

To enumerate all the sculptors this century who have tried their hand at bronze would be tantamount to compiling a "Who's Who in Sculpture 1900–1972," so many are the great and less great who have cast at least one work during their lifetime. In any case such an attempt would fall outside the limits of this book which is dedicated to the founder's art, not the sculptor's.

In this connexion mention should be made once again of Miguel Berrocal, the Spanish founder now living in Italy, who has introduced industrial casting processes to the art world in order to satisfy a demand for large editions of small works of art, not in the 19th century way, but signed and numbered like a silkscreen print or a lithograph.

The quality and quantity of 20th century bronzes, whatever the technique used to produce them, provide ample testimony of the public's renewed interest in this form of expression and of the prominent position accorded today to art foundry.

SYNOPSIS

Art foundry	The Arts	General
5000 B.C. First known bronze	Stone engraving, ceramics, pottery	End of polished stone or Neolithic period
2500 B.C. Casting technique spreads north from Near East, to the Caucasus, Egypt, Crete	Sumerian funeral architecture. Egyptian painting and sculpture at its zenith	Aryan invasions
2000 to 1500 B.C. Tools and weapons in bronze in general use	Zenith of Cretan art	Egypt: Middle Empire, Thebes
1000 B.C. First bronzes cast in Luristan (Iran) First Egyptian bronzes	Hebrew music, metalwork, Egyptian animal art	David, king of Israel
650 B.C. "Karlsruhe" female head, first large scale hollow casting	Etruscan art Greece: Ionic Order	
650–525 B.C. Egyptian art foundry at its zenith during the Saite Dynasty	Greece: Sapho's poetry, temple of Apollo, Delphi	Nebuchadnezzar in Jerusalem
500–100 B.C. Zenith of Greek sculpture and bronze. Myron, Polycletus, Phidias, Praxiteles, etc. *Discobolos, Apollo Belvedere, Venus of Milo,* etc.	Zenith of Persian art Art of the Huns	Zenith of Greek civilization Alexander the Great conquers Asia Minor

Art foundry	The Arts	General

1st thousand years A.D.
    Casting used in Roman, early
    Christian, Chinese, Byzantine,
    Burgundian arts
	Roman theatre and literature	Roman Empire, Franks,
		Charlemagne Roman Emperor
		Germanic Holy Roman
		Empire

11th century
    First *Ife* bronzes (Benin region)
    Late 11th century–late 12th century
    Doors of S. Zeno basilica Verona
	Zenith of Romanesque art
	Chinese paintings
	Birth of Gothic art
		William the Conqueror
		First Crusade
		Philippe-Auguste 1180–1223

1330 Andrea Pisano: first baptistry doors in Florence	Zenith of Gothic art Giotto and the Italian Primitives	Gengis-Khan invades China Hundred Years' War
1378 Birth of Lorenzo Ghiberti		Charles V and Duguesclin recover land ceded to the English
1386 Birth of Donatello	Birth of Fra Angelico Birth of Jan van Eyck Birth of Paolo Uccello	

1400 Competition for the decoration
       of the second portal of the
       Florence baptistry
	Birth of
	Roger van der Weyden
	Birth of Conrad Witz	

1402– 1452 Ghiberti: second and third doors of the Florence baptistry	End of the Middle Ages Beginning of the Renaissance	Jean sans Peur, duke of Burgundy
1409 Donatello: David	Gentile da Fabriano and Pisanello decorate the Doges' palace Flemish school: van Eyck, van der Weyden, Memling	Charles VII, king of France

Art foundry	The Arts	General
1429		Joan of Arc enters Orleans
1432 Birth of Antonio Pollaiuolo		
1434	Florentine school: Masaccio, Fra Angelico, Donatello, Lippi, Uccello, Verrocchio, etc.	Cosimo de' Medici, the Elder, chief of Florentine Republic
1452 Birth of Leonardo da Vinci		Charles le Téméraire, duke of Burgundy
1466 Death of Donatello		
1469		Lorenzo de' Medici "The Magnificent" rules Florence
1475	Birth of Michelangelo Venetian school: Carpaccio, the Bellinis, Verocchio (*Colleone* monument)	
1483	Birth of Raphael	Columbus discovers America
1500	German school: Dürer, Holbein, Cranach, Grünewald	Alexander VI Borgia, Pope
1512 Maximilian I creates a foundry near Innsbruck to cast statues for his tomb	Mathias Grünewald: *Isenheim retable*	
1513		Leo X pope
1515 Riccio (Andrea Briosco): paschal chandelier at St. Anthony's basilica, Padua	Leonardo da Vinci in France	François I, king of France
1519	Death of Hieronymus Bosch	Birth of Cosimo I de' Medici Catherine de Médicis Charles V, emperor

116

Art foundry	The Arts	General
1524 Birth of Giovanni Bologna 1527 Jacopo Sansovino installed at Venice	Loire castles Fontainebleau school	Sack of Rome by Charles V's troops
1537	Venetian school: Titian, Veronese, Tintoretto, etc.	Cosimo I ruler in Florence Ignatius Loyola: The Society of Jesus
1554 Benvenuto Cellini: Perseus	Breughel: *Massacre of the Innocents* Goudimel: *Psalms*	Mary Tudor, queen of England
1571 Death of Cellini	Birth of Rubens	Massacre of St. Bartholomew
1580 First assembled bronzes		William of Orange Henry IV, king of France
1598 Birth of Bernini	Birth of Mansart Decadent Italian period: Caravaggio, Carracci, Gherardi	Edict of Nantes Marie de Médicis, regent
1624 Death of Antonio Susini, the founder in whose workshops most of Giovanni Bologna's work was cast	Dutch school: Rembrandt, Frans Hals, Vermeer, etc.	Richelieu minister
1628 Birth of Girardon, author of the equestrian statue of Louis XIV	Rembrandt: *The Anatomy Lesson* Flemish school: Breughel, Rubens, Van Dyck, etc.	Louis XIV, king of France
1650 Bronze used in cabinet-making, as exploited later by Boulle		
1665 Bernini presented to Louis XIV at Saint-Germain	Versailles gardens Le Brun: Louvre decorations	

Art foundry	The Arts	General
1680 Death of Bernini Keller brothers cast the equestrian statue of Louis XIV		Zenith of Louis XIV's power
1684	Birth of Watteau	Revocation of the Edict of Nantes
1728 Engraving by J.A. Meissounier published which launches the "rococo" bronze style	Chardin: *La Raie* Bouchardon: rue de Grenelle fountain	Fleury minister Peter II, czar
1741 Birth of Houdon		Elizabeth, czarina
1772 Pitoin, Louis XV's bronze founder, casts a pair of fire-dogs for Mme du Barry	Piranese: prints Japanese painters and printmakers: Hokusai, Outamaro	Seven Years' War
1776 In Paris the guilds of the founder- chasers and the chiseler-gilders amalgamate	Birth of Turner	George Washington
1784 Birth of François Rude	David: *Les Horaces*	Washington founded
1792 Birth of Pradier	Birth of Géricault	Republic proclaimed in France
1796 Birth of Antoine-Louis Barye	Birth of Corot	Napoleon in Italy
1807 François Rude installed at Paris	Beethoven: 5th Symphony	Battle of Eylau
1808 Birth of Daumier		
1818 French founders form association of bronze foundry-men and employ artists	Géricault: *Le Radeau de la Méduse*	Bolivar frees Columbia and Chile
1827 Birth of Jean-Baptiste Carpeaux	Corot: *Le Pont de Narni*	Nicholas I, czar
1828 Death of Houdon	Birth of Bœcklin	
1833 Rude: *La Marseillaise*		
1839 Collas invents a machine for reproducing an original model on any scale	Birth of Cézanne	Guizot ministry

Art foundry	The Arts	General
1840 Barye: *Le Lion de la Bastille* Birth of Rodin	Birth of Monet	
1848	Birth of Gaugin	Revolution in Europe
1861 Births of Maillol and Bourdelle		
1870	Cézanne: *L'Estaque*	
1876 Birth of Constantin Brancusi	Renoir: *Le Moulin de la Galette*	German Empire, William I
1879 Death of Daumier	Impressionism	
1880 Rodin: *Le Penseur*	Death of Manet	Alexander III, czar
1885 Birth of Henri Laurens		
1886	Seurat: *La Grande Jatte* Van Gogh: *The Sunflowers*	
1890 First casting of Daumier's bronzes	Death of Van Gogh	2nd Internationale
1895 Rodin: *Les Bourgeois de Calais*		
1898 Birth of Henry Moore		Dreyfus affair in France
1900 Paris Exhibition, a whole pavilion is devoted to Rodin	Ravel: *Jeux d'eau* Ecole de Paris	Boer War
1908 Brancusi: *Le Baiser*	The Fauves Matisse: *Les Poissons rouges* Cubism	Revolution in Turkey
1912 Maillol: *Pomone*		War in Balkans
1918 Death of Duchamp-Villon	Stravinsky, Ramuz, Auberjonois: *L'Histoire du Soldat*	Russian Revolution
1921 Pompon: *Ours*		
1922 Birth of Joannis Avramidis		Fascism in Italy
1923 Archipenko opens an art school in New York	Deaths of Monet and Valloton	Stalin assumes power

Year	Art foundry	The Arts	General
1929	Freundlich: Monumental abstract sculpture	Le Corbusier: Cooperatives' palace, Moscow	World-wide economic crisis
1930	Lipchitz exhibits 100 works in Paris		
1931	First sculpture in the round by Jean Arp		Germany: Nazi party
1930–1932	Gonzales and Picasso work together	Bonnard: *Grand nu au miroir* Matisse: *La Danse*	Hitler's rise to power
1939	Rodin's *Balzac,* refused in 1898 by *La Société des gens de lettres* is at last erected in Paris	Rouault: *Christ aux outrages*	2nd World War
1942	Death of Gonzales	French school: Estève, Pignon, Bazaine, Brianchon	
1944	Death of Maillol		Allied landings in Normandy
1946	Henry Moore retrospective exhibition Museum of Modern Art, New York	Rebirth of stained glass: Rouault, Cingria, Monnier Death of Bonnard	U.N.O.
1948	Pevsner-Gabo exhibition, Musée d'Art moderne, Paris	Chagall returns to France	IVth Republic in France State of Israel created
1953	Zadkine: commemorative monument, Rotterdam		Korean War
1954	Death of Laurens	Tachisme Death of Matisse Death of Nicolas de Staël	Dien Bien Phu
1956	Gonzales exhibition at the Museum of Modern Art, New York Arp: big bronze relief for UNESCO, Paris	Death of Pollock	Hungarian rising

Art foundry	*The Arts*	*General*
1957 Before his death Brancusi acquires French nationality and bequeaths his studio to the state on condition that it become a museum Gabo: steel monument for Rotterdam		Treaty of Rome: Common market 5th Republic in France Khrushchev in U.S.S.R.
1961 Archipenko: *The Queen of Sheeba,* his last bronze – 165 cm high	Pop Art in U.S.A.	Berlin Wall
1962	Death of Braque	First space flight
1966 Berrocal's first multiples Death of Giacometti		War in Viet-nam
1967–1969 Travelling retrospective exhibition of Archipenko's work organized by the Smithsonian Institute		

GLOSSARY OF TECHNICAL TERMS

Chase. To work the surface of a metal (with chisel, engraving tools, matting tools, burin, etc.).

Core. An interior mould necessary to cast a hollow statue.

Crucible. Container made of graphite and carborundum in which the metal is melted.

Dross. Impurities on the surface of molten metal.

Freezing. Said of liquid wax or molten metal when it starts to solidify.

Furnace. Term which applies both to the construction of fireproof material built round the moulds when the wax is melted out, and also to the bronze-melting furnace.

Gelatine. Term usually applied to the negative mould into which liquid wax is poured. This mould may be in gelatine, but also in elastomere, or latex, or other silicon based substances.

Grog. Pulverized ceramic piping.

Investment. Strong, plastic, fireproof, porous material used for building the ludo mould round the waxes, and also for the core. Usually made from a mixture of plaster of Paris, brick powder and grog.

Ludo mould. Fireproof mould built round the waxes.

Master cast. The artist's original work from which the founder makes waxes which are melted out of the ludo mould (lost wax), and replaced by bronze. In lost wax casting the master cast can be in almost any material and is not destroyed.

Matting. To give a mat appearance to the surface of a bronze. Also, to work over a correction on a bronze so that it resemble the surrounding surface.

Mould. Negative imprint of an original work, can be in plaster, sand, gelatine, etc.

Noble metals. Copper, tin, bronze, silver, gold and platinum.

Pantograph. An instrument for reproducing a work mechanically to any scale.

Patina. Colouring of metal which either occurs naturally, or is obtained by chemical means.

Pouring gate. Orifice in ludo mould through which the molten metal is poured before spreading through the system of runners.

Risers. Vents through which air and gases escape when molten metal is poured into the ludo mould.

Runners. Channels through which the molten metal flows into the ludo mould to fill the impression left by the lost wax.

Runner cup. See pouring gate.

Shank and ring. Metal ring held by two shafts or handles one of which ends in a T, in which is placed the crucible containing molten metal. It is supported by two men, one of whom turns the T-bar to pour the metal into the moulds.

Slag. See Dross.

Tamp. To pack down (sand in sand casting).

Tongs. Large tongs are used in casting to move the crucible of molten metal from the furnace into the shank and ring, also to remove hot castings from sand moulds, and even to add hot metal to the furnace.

Wax. Term applied to the positive wax model made from a master cast.

Waxes. The waxes used for casting are usually a mixture of rosin, bee's wax, turpentine, paraffin wax and ceresin.

SELECT BIBLIOGRAPHY

Charbonneaux, Jean, *Les Bronzes grecs.*
  Paris, 1958.
Chastel, André, *Le Grand Atelier d'Italie.*
  Paris, 1965.
Landais, Hubert, *Les Bronzes italiens de la Renaissance.*
  Paris, 1958.
Mills, John W., and Gillespie, Michael, *Studio Bronze Casting – lost wax.*
  London, 1969.
Montagu, Jennifer, *Les Bronzes.*
  Paris, 1965.
Seuphor, Michel, *La Sculpture de ce Siècle.*
  Neuchâtel, 1959.

TABLE OF ILLUSTRATIONS

PHOTOGRAPHIC CREDITS

Galerie Claude Bernard, Paris: p. 59, 98, 104.
Photo Eliane Bouvier, Geneva: cover, p. 2, 13, 16, 19, 24, 25, 28, 29, 30, 31, 34, 35, 36, 37, 38, 40, 41, 42, 43, 44, 45, 46, 47, 50, 51, 52, 53, 55, 56, 57, 58, 59, 60, 61, 62, 64, 66, 67, 68, 69, 70.
Photo Nicolas Bouvier, Geneva: p. 6, 15, 16, 21, 49, 65.
Photo Gad Borel-Boissonnas, Geneva: p. 87.
Gimpel and Hanover Gallery, Zurich: p. 111.
Photographies et clichés Giraudon, Paris: p. 62, 77 (4), 86, 88, 94, 96, 99, 101, 103, 105.
Photos Grivel, Geneva: p. 92, 106, 107.
Christian Hauser, Geneva: p. 85.
Photo André Held, Lausanne: p. 39, 76, 78, 79, 81, 82, 83, 93, 97, 102.
Galerie Krugier & Cie, Geneva; photo Claude Mercier: p. 8, 26, 33, 112.
Claude Mercier, Geneva: p. 72, 73.
Musée d'Art et d'Histoire, Geneva; photo Yves Siza: p. 89.
Photo Musée d'ethnographie, Geneva: p. 54.
Clichés des Musées nationaux, Paris: p. 74, 77 (1, 2, 3), 95, 108, 109, 111.
Fred Pillonel, Geneva: p. 18, 38.
Photo Ernst Scheidegger, Zurich: p. 22.
Yves Siza, Geneva: p. 9, 10, 11.

The author and publisher wish to express their thanks to those who helped work on the preparation of this book; in particular Jean-Marie Pastori, the Geneva lost wax art founder and his associates, Jean-Claude Reussner and Jean Donzé, sand cast founders at Fleurier/Neuchâtel, who kindly made their workshops available for the technical photography, equally to Jean-Pierre Durand, Geneva; the museum staffs who put their documents at the author's disposal, and to the following galleries: Claude Bernard, Paris; Maeght, Paris; Krugier & Cie, Geneva.

English version by Julian Snelling and Claude Namy.

Printed in Switzerland